Solar System

Grades 2-3

by
Laura Phou

Frank Schaffer Publications®

Table of Contents

Author: Laura Phou
Cover Illustration: Ron Lipking

Frank Schaffer Publications®

Send all inquiries to:
Frank Schaffer Publications
8720 Orion Place
Columbus, Ohio 43240-2111

Solar System—Grades 2-3

ISBN: 0-7682-0508-5

8 9 10 11 12 GLO 12 11 10 09

What's in Our Solar System?

An *astronomer*, or scientist who studies the universe, might make this list if you asked her what is in our solar system.

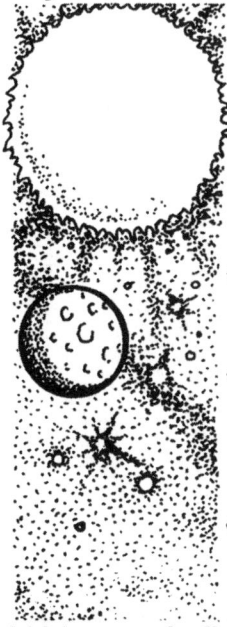

- one **star**, or hot glowing ball of gases, called the Sun

- all the planets' moons

- small chunks of rock or ice called **meteoroids**

- lots of empty space

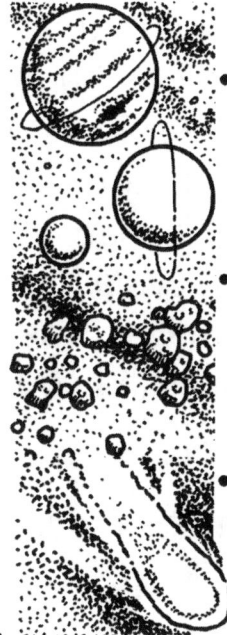

- nine worlds called **planets** that travel around the Sun

- chunks of rock and metal called **asteroids**

- frozen balls of dirty ice called **comets**

Write a definition for each of these words.

1. astronomer _____

_____

2. star _____

3. planets _____

_____

4. asteroids _____

5. meteoroids _____

_____

6. comets _____

_____

Brainwork! Find *solar system* in a dictionary or in the glossary of a science book. Write the definition you find.

Our Sun

When you see the Sun shining during the day, you are seeing a star. A star is a huge glowing ball of gases. The Sun is the only star in our solar system. It looks much larger than the stars we see at night because it is closer to us than the others. Even so, the Sun is 93 million miles from Earth.

Our Sun is really only a medium-size star. Some other stars in the universe are much bigger, and many stars are much smaller. The Sun is a yellow star. Hotter stars are blue and cooler stars are red.

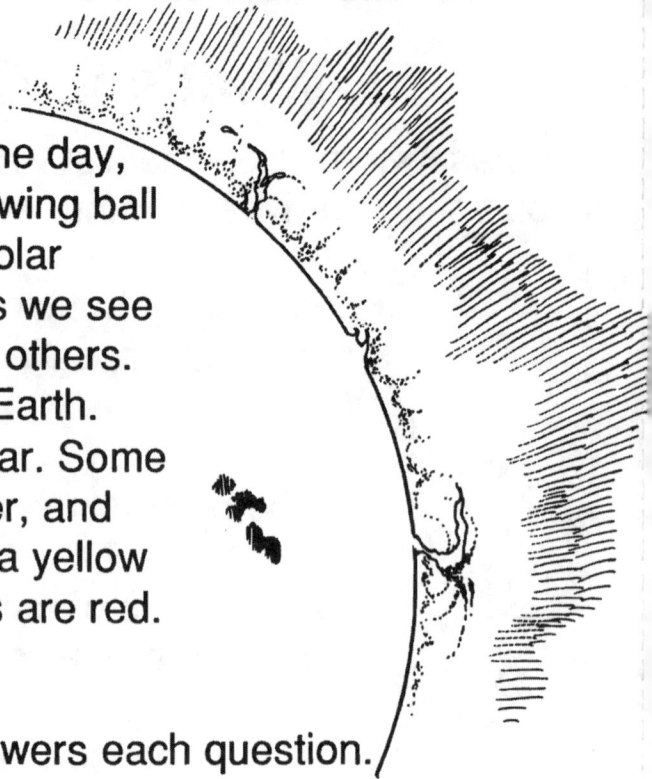

Copy the sentence from the story that answers each question.

1. What is a star? _____

   _____

2. Which star is in our solar system? _____

   _____

3. How far is the Sun from Earth? _____

   _____

4. What color is the Sun? _____

5. Why does the Sun look larger to us than other stars?

   _____

   _____

   _____

Brainwork! The Sun's light and heat help Earth's plants and animals to grow. Draw a picture to show this.

Name _____

The Planets Are Moving!

Each of the planets in our solar system **revolves**, or travels, around the Sun. The planets circle the Sun along paths called **orbits**. Because the planets are at different distances from the Sun, each one takes a different length of time to revolve once.

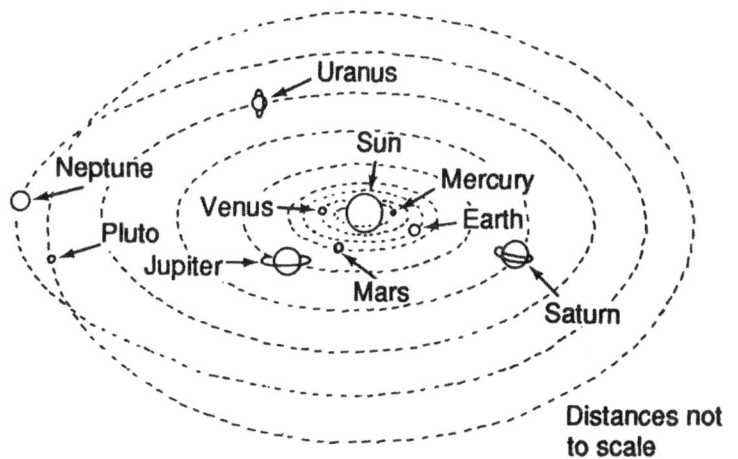

Distances not to scale

1. What word means *travels around*? _____

2. What are the planets' paths around the sun called? _____

3. Why do the planets take different lengths of time to revolve around the

    Sun?_____

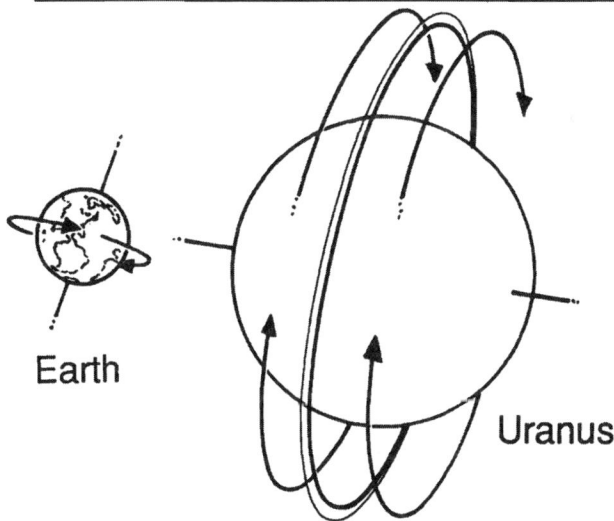

Earth

Uranus

Each planet in our solar system **rotates**, or **spins**, around a line through its center. This imaginary line is called an **axis**. It takes each planet a different length of time to rotate once.

4. Each planet _____ around a line through its center.

5. This imaginary line is called an _____.

6. *Rotates* means _____.

Brainwork! Use two things from your desk. Move one so it revolves around the other. Then put one down and move the other so it rotates.

Solar System Scramble

Unscramble the name of each numbered object below. Write the name on the correct line below.

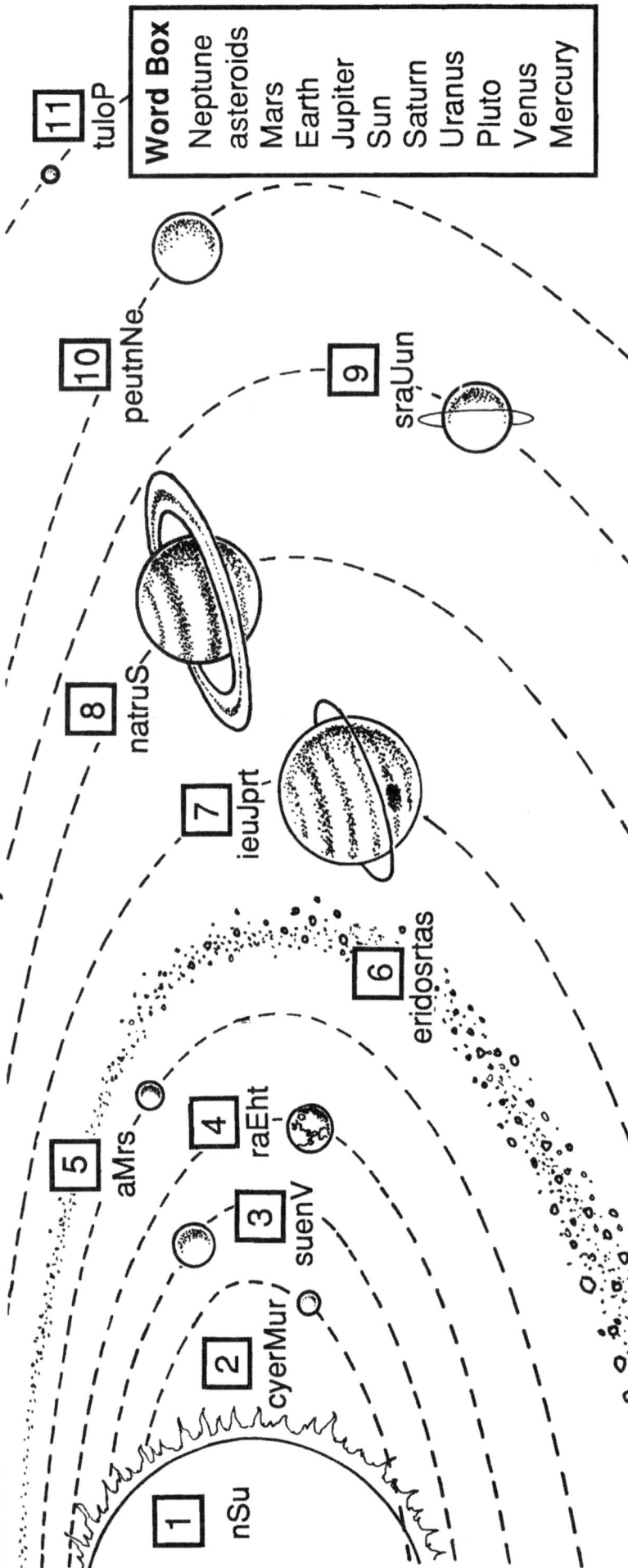

Word Box
Neptune
asteroids
Mars
Earth
Jupiter
Sun
Saturn
Uranus
Pluto
Venus
Mercury

1. nSu

2. cyerMur

3. suenV

4. raEht

5. aMrs

6. eridosrtas

7. ieuJprt

8. natruS

9. sraUun

10. peutnNe

11. tuloP

1. _____
2. _____
3. _____
4. _____
5. _____
6. _____
7. _____
8. _____
9. _____
10. _____
11. _____

Brainwork! Turn this paper over and write the names of the nine planets in our solar system.

4

FS-8510 Solar System

A Strip of Space

Follow these directions to compare the positions of the planets from the Sun.

1. Color:

- the Sun yellow
- Mercury brown
- Venus yellow

- Earth green
- Mars red
- Jupiter orange

- Saturn yellow
- Uranus and Neptune blue
- Pluto purple

2. Cut out the four strips.

3. Glue:

- strip 2 to the right end of strip 1
- strip 3 to the right end of strip 2
- strip 4 to the right end of strip 3

(Distances are to approximate scale.)

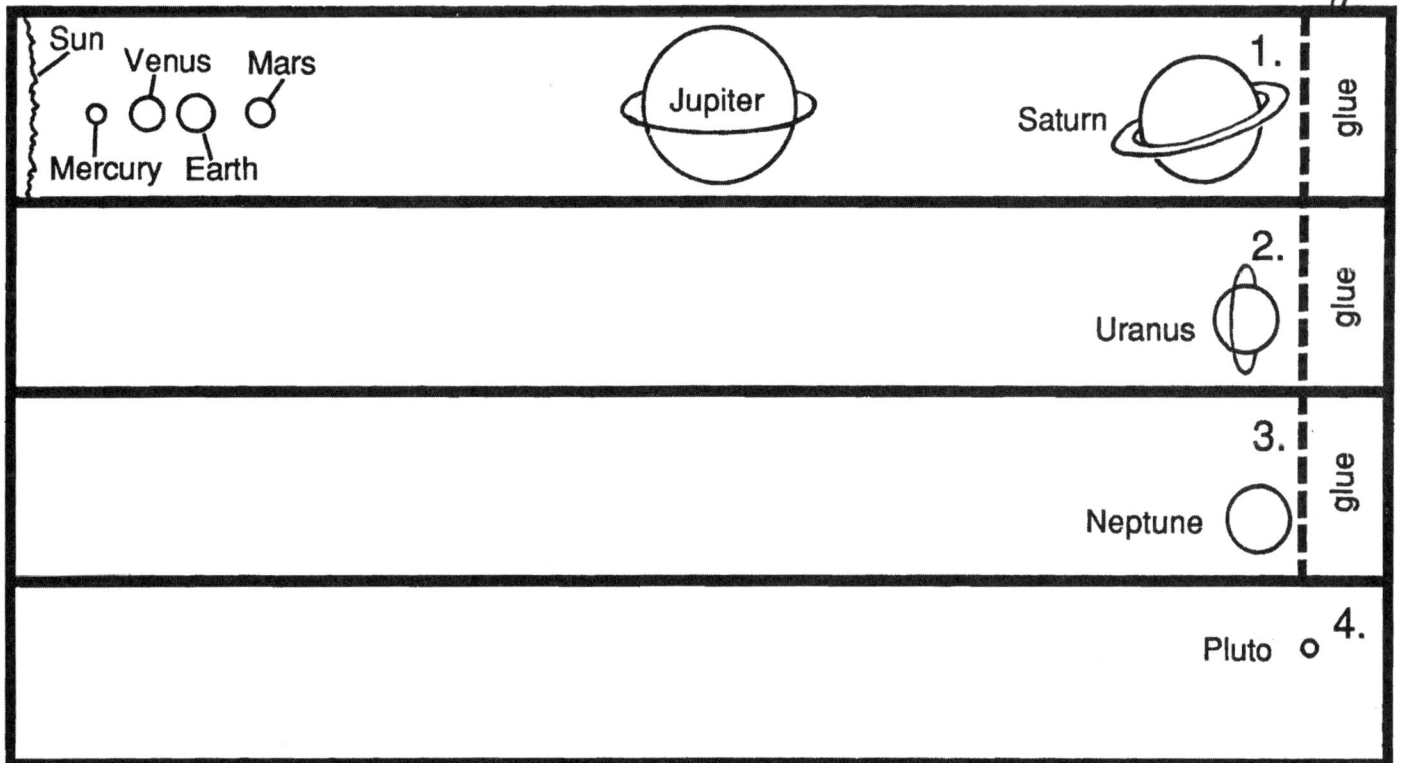

Sun   Venus   Mars   Jupiter   Saturn   1.   glue
Mercury   Earth

2.   Uranus   glue

3.   Neptune   glue

Pluto   4.

Mercury—Closest to the Sun

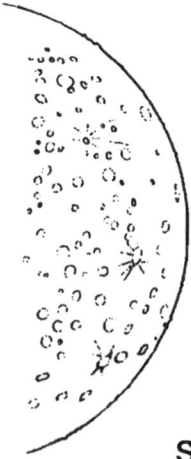

Mercury is the planet closest to the Sun. That is why Mercury travels around the Sun faster than any other planet. It takes Mercury 88 days to revolve once around the Sun.

Little was known about Mercury before 1974. Scientists have a hard time studying Mercury with telescopes because of the Sun's great light. In 1974 and 1975 an unmanned spacecraft named *Mariner X* flew by Mercury three times and sent scientists new information about the planet.

The surface of Mercury is much like the moon's surface. It has high cliffs and deep craters, or holes. Mercury has almost no atmosphere, or gases surrounding it. Temperatures on the planet range from 950° F to −210° F! Mercury has no moons.

Write each answer in a sentence.

1. Which planet is closest to the Sun? _____

_____

2. How long does it take Mercury to revolve around the Sun? _____

_____

3. Why do scientists have a hard time studying Mercury with telescopes?

_____

4. What did Mariner X do? _____

_____

_____

5. Describe Mercury's surface. _____

_____

Brainwork! Make a list of three interesting facts about Mercury.

Venus—Earth's Twin

Use the words in the Word Bank to complete the story.

Word Bank			
light	against	lightning	size
closest	higher	atmosphere	melt

Venus has been called Earth's twin because it is about the same _____ as Earth. Venus is the second
1
planet from the Sun and is the planet _____ to
2
Earth. Venus was also the first planet to be studied by spacecraft. Venus has no moon.

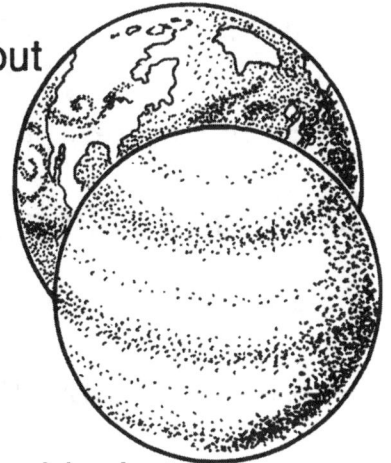

Venus has an interesting _____ , or blanket
3
of gases around it. It reflects, or bounces off, so much of the Sun's _____ that Venus is easier to see than any other planet. The
4
atmosphere also lets some sunlight in and traps heat _____ the
5
planet's surface. Therefore, temperatures on Venus are high enough to _____ some metals. Clouds move at high speeds in Venus's
6
atmosphere, and bolts of _____ streak across its sky.
7

Venus has volcanoes on its surface and a mountain _____
8
than the highest on Earth. There is no liquid water on Venus. Earth's plants and animals could not live on Venus.

Brainwork! Think of another nickname for Venus. Write to tell why it is a good nickname.

Our Home Planet

Use the words from the Word Bank to complete the story.

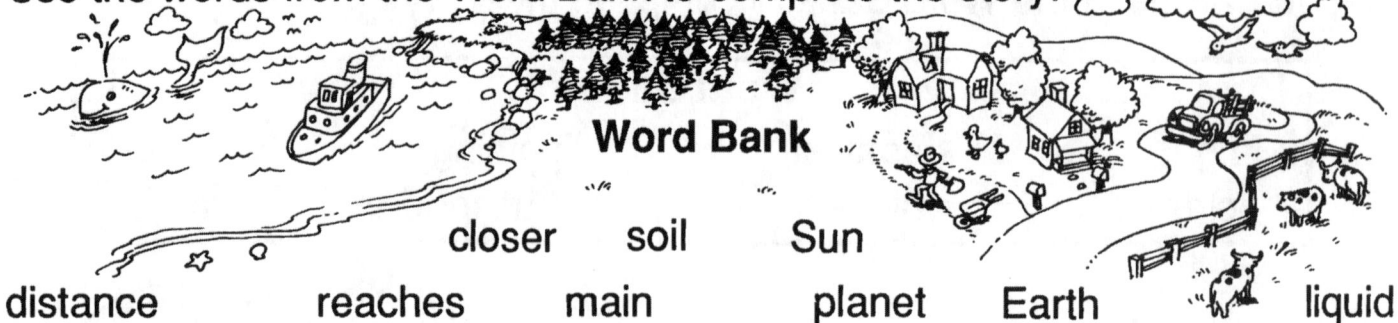

Word Bank

closer   soil   Sun

distance   reaches   main   planet   Earth   liquid

The third planet from the _____ is our home planet Earth. Earth has
1

something no other _____ is known to have—living things.
2

Earth is at the right _____ from the Sun to have the liquid water
3

necessary to support life. Mercury and Venus are too hot because they are

_____ to the Sun. The other planets are too far from the Sun to
4

have _____ water. Not much heat or light _____ them, so the
5   6

water would be in the form of ice.

Earth has a lot of water. Most living things need water. Water helps to

control the earth's weather and climate. Water also breaks rocks into

_____ which plants need to grow.
7

Earth is surrounded by a blanket of air called the atmosphere. Oxygen is

one of the _____ gases in the atmosphere. Most animals breathe
8

oxygen.

_____ is a special planet!
9

Brainwork! Design a poster showing why Earth is a good planet for
living things.

We See Our Moon

Earth has one moon. It is the moon that we see in the sky. The moon is Earth's partner in space. It makes a path around, or **orbits**, Earth. It also orbits the Sun along with the earth.

The moon looks large because it is closer to Earth than the Sun or planets. Four moons would stretch across the **diameter**, or widest part of the earth.

In 1969 **astronaut** Neil Armstrong took the first steps on the moon. Scientists have studied rocks brought back from the moon.

The surface of the moon has many deep holes called **craters**. It has flat areas called **maria**. The moon also has rocky mountain areas called **highlands**. There is no air, wind, or water on the moon. No life exists there.

Write the word in dark print from the story that matches each definition.

1. deep holes in the moon's surface

2. to make a path around

3. flat land on the moon

_____

4. the widest part of the earth

5. areas with rocky mountains

6. a person who travels in space

_____

Write two sentences about the moon using two of the words in dark print.

1. _____

_____

2. _____

_____

Brainwork! Would you like to visit the moon? Write to explain your answer.

Mars—The Red Planet

Mars, the fourth planet from the Sun, is half the size of Earth. Mars has two moons. It has been called the Red Planet because of its red color. Parts of this planet's surface are covered with sand dunes and dry reddish deserts. Other areas look like dried up riverbeds. Some scientists believe water may once have flowed on Mars. Mars also has two polar caps made up of frozen water and dry ice. Pink, blue, and white clouds move through the Red Planet's sky.

For a long time some people thought there might be life on Mars. When two U.S. spacecraft landed on the planet in 1976, they sent back photographs of Mars and did experiments to find out if life exists there. Scientists now believe that Mars does not have plant or animal life like that on Earth.

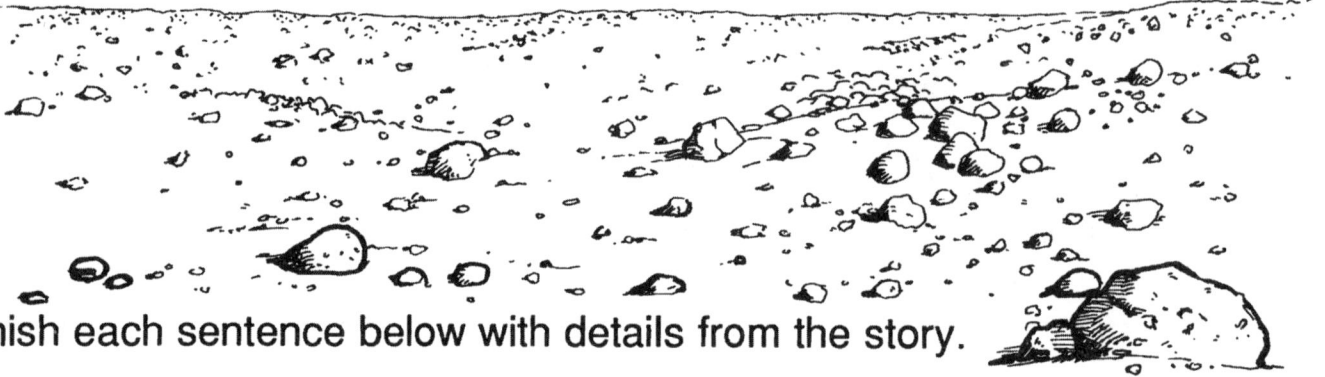

Finish each sentence below with details from the story.

1. Mars is the _____ planet from the Sun, and it has _____ moons.

2. Mars is nicknamed the _____ .

3. Two U.S. spacecraft landed on Mars in _____ , sent back photographs, and did _____ .

4. Mars has dry reddish _____ and what look like dried up _____ .

5. Mars has two _____ made of frozen water and dry ice.

Brainwork! Draw and color a picture that shows your idea of the Red Planet's landscape.

Jumbo Jupiter

Jupiter is the largest of the nine planets. It is more than 11 times larger than Earth.

Jupiter is the fifth planet from the Sun, and it travels once around the Sun every 12 years. This jumbo planet rotates in just ten hours—faster than any other planet!

Thick clouds surround Jupiter. Most scientists believe that the belts of color in Jupiter's atmosphere are caused by different gases. The planet is a giant ball of liquids and gases with, perhaps, a small rocky core. Its famous Great Red Spot is a huge storm of swirling gases. Lightning streaks across Jupiter's sky. Jupiter has a thin dust ring around its middle and 16 known moons.

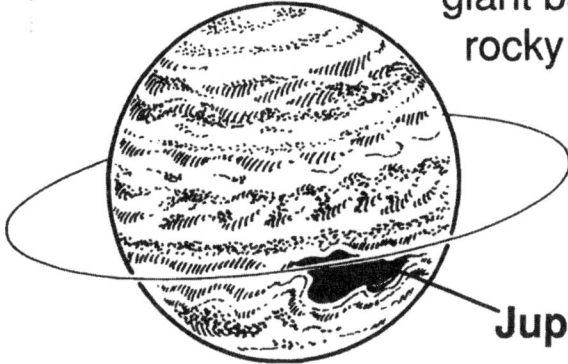

Jupiter's Great Red Spot

Write **true** or **false**.

_____ 1. Jupiter is the smallest planet in our solar system.

_____ 2. Earth is larger than Jupiter.

_____ 3. It takes 12 years for Jupiter to travel around the Sun.

_____ 4. Jupiter rotates faster than any other planet.

_____ 5. Jupiter's Great Red Spot is a huge storm of swirling gases.

_____ 6. Jupiter has a thick ice ring around its middle.

_____ 7. Jupiter has more than ten moons.

_____ 8. Jupiter is the sixth planet from the sun.

Brainwork! Write one true and one false statement about Jupiter. Have a friend tell which is true and which is false.

Stunning Saturn

Saturn is the sixth planet from the Sun. Saturn is best known for the beautiful rings around its middle. The rings are thin and flat and made of pieces of rock and ice. They stretch more than 100,000 miles across!

Some scientists believe the rings are made of particles left over from the time when Saturn first became a planet. Others believe the rings are made of pieces of a moon that was torn apart when it came too close to Saturn.

Saturn is the second largest planet. Since Saturn is more than nine times farther than Earth is from the Sun, it is much colder than Earth. The planet is a giant ball of spinning gases. Saturn has at least 20 moons.

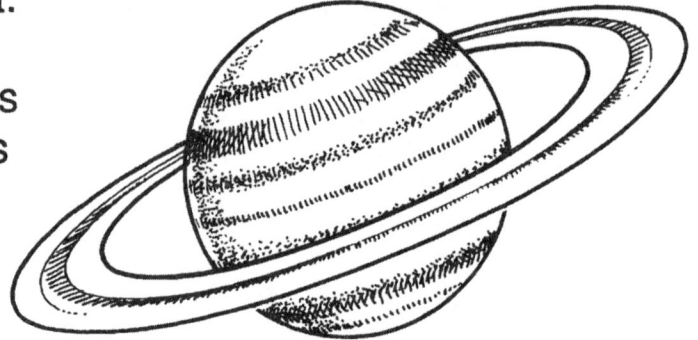

Write each answer in a sentence.

1. For what is Saturn best known? _____

_____

2. What is one idea scientists have about how Saturn's rings were made?

_____

_____

3. How does Saturn compare in size with the other planets? _____

_____

4. Why is Saturn colder than Earth? _____

_____

5. How many moons does Saturn have? _____

_____

Brainwork! Write a poem about Saturn's beautiful rings.

The Blue-green Giants

Uranus and Neptune are giant planets more than a billion miles from the Sun and Earth. They are about the same size. Each is more than $3\frac{1}{2}$ times larger than Earth. They look blue-green in photos because both have a gas called methane in their atmospheres. Uranus and Neptune are very cold planets where life probably doesn't exist.

Uranus is the seventh planet from the Sun. It is known to have at least 15 moons and 11 thin rings. Uranus rotates in the direction opposite to that of Earth. It can be seen from Earth without a telescope.

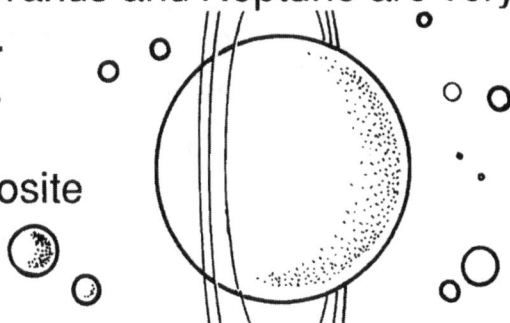

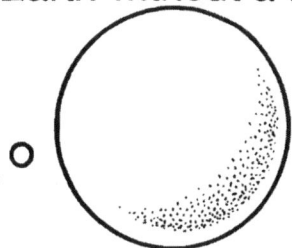

Neptune is farther from the Sun than Uranus. It has eight known moons. Some astronomers believe it may also have a ring. Neptune cannot be seen without a telescope.

Decide which planet or planets each fact describes. If it describes Uranus, write *Uranus*. If it describes Neptune, write *Neptune*. If it describes both Uranus and Neptune, write *both*.

1. rotates in the opposite direction

_____

2. called a blue-green giant

_____

3. cannot be seen without a telescope

_____

4. is more than a billion miles from Earth

_____

5. has methane in its atmosphere

_____

6. has at least 11 rings

_____

7. can be seen without a telescope

_____

8. has eight known moons

_____

Brainwork! List three ways Uranus and Neptune are alike. List three ways they are different.

Faraway Pluto

Pluto travels farther from the Sun than any other planet in our solar system. At its farthest point, it is more than four billion miles from Earth!

Pluto is also the smallest of the nine known planets. It is smaller than Earth's moon.

Scientists know very little about the planet Pluto because it is so far away. It is believed to be like a rocky snowball in space. Charon is Pluto's only moon. Scientists don't think any life exists on faraway Pluto.

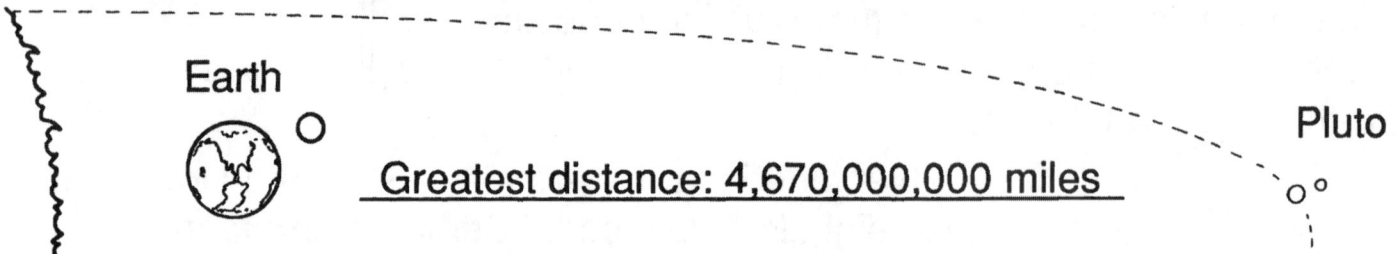

Earth

Pluto

Greatest distance: 4,670,000,000 miles

Unscramble each sentence so it tells one fact about Pluto. Write the fact.

1. farthest Sun Pluto travels from the

   _____

2. has moon one Pluto

   _____

3. planet smallest Pluto is the

   _____

4. travels billion more four than Earth from miles Pluto

   _____

   _____

5. Pluto's named is Charon moon

   _____

Brainwork! Write two questions you would like to ask an astronomer about Pluto or its moon.

So Far Apart

Mercury
Venus
Jupiter
Earth
Mars

Saturn

Uranus

Neptune

Pluto

Sun

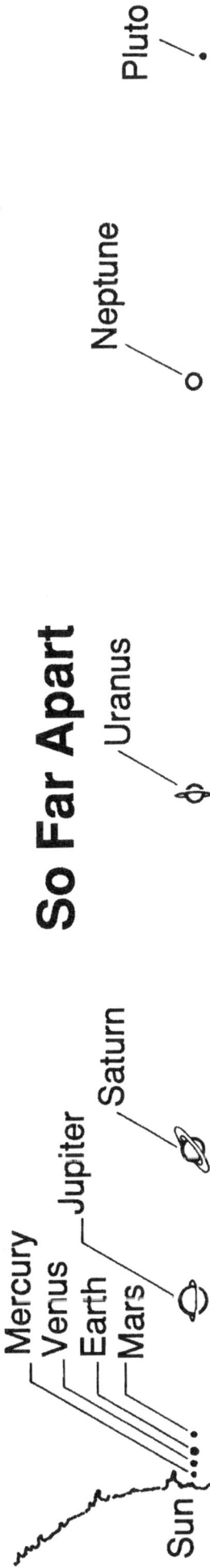

Sizes of planets not to scale

Planets and Their Average Distances From the Sun

Mercury	36 million miles	Mars	142 million miles	Uranus	1,781 million miles
Venus	67 million miles	Jupiter	484 million miles	Neptune	2,788 million miles
Earth	93 million miles	Saturn	885 million miles	Pluto	3,660 million miles

Use the chart and diagram to answer these questions.

1. What is Neptune's average distance from the Sun?

2. Which planet has an average distance from the Sun of 142 million miles?

3. Which planet is closest to the Sun—Saturn, Mars, or Neptune?

4. How much farther from the Sun is Venus than Mercury?

5. How much farther is the fourth planet from the Sun than the third planet from the Sun?

Brainwork! Write why you think it is difficult for people to travel to other planets.

The Planets' Names

Match each symbol in the puzzle to a clue below. Write the planet's name across or down in capital letters.

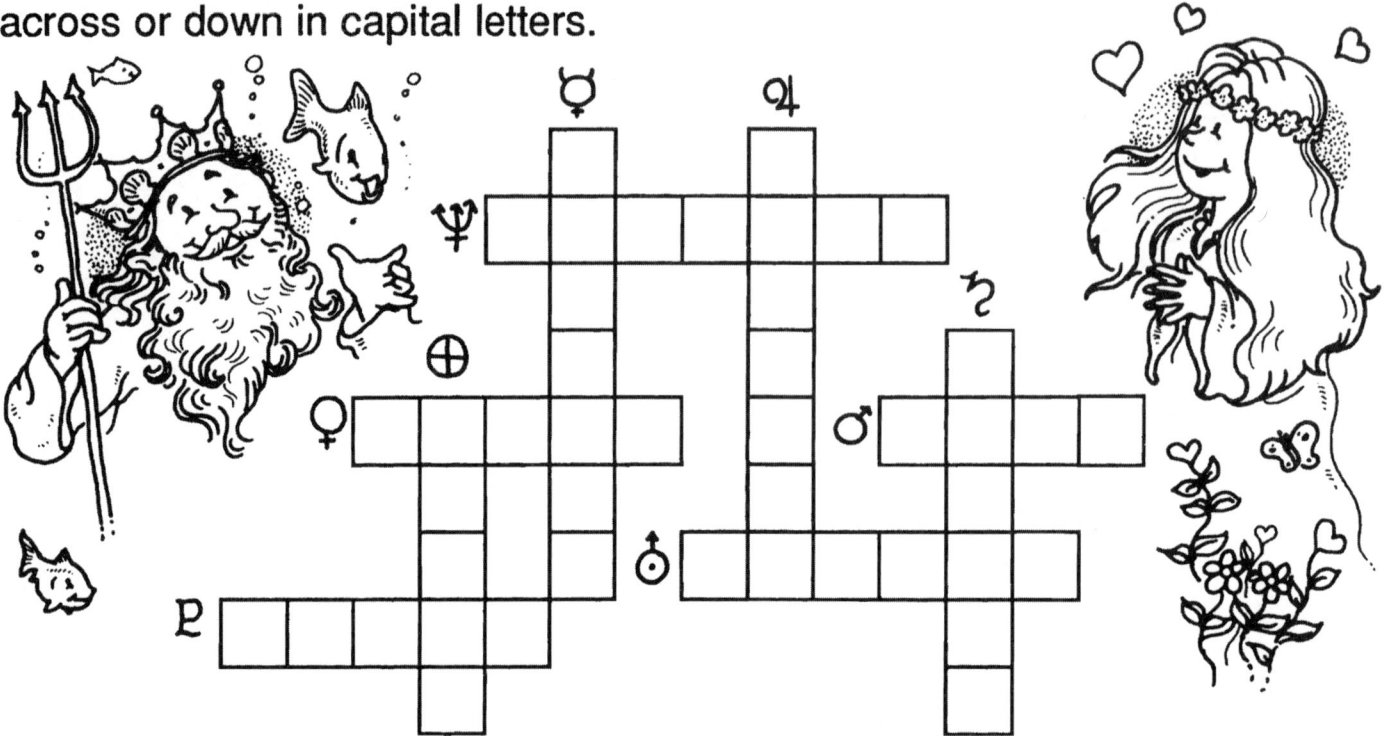

Across

⍦ Neptune was named after the Roman god of the sea.

♀ Venus was named after the Roman goddess of love and beauty.

♂ Mars was named after the Roman god of war.

♅ Uranus was named after the Greek god of the sky.

♇ Pluto was named after the Greek and Roman god of the lower world.

Down

☿ Mercury was named after the Roman messenger of the gods.

♃ Jupiter was named after the Roman king of the gods and ruler of the universe.

♄ Saturn was named after the Roman god of farming.

♁ Earth was named after the Greek earth goddess.

Brainwork! Make a word search puzzle with the planets' names. Have a friend solve your puzzle.

My Planet Report

Name of the planet _____

Named after _____

Size of the planet _____

Average distance from the Sun _____

Time needed to revolve around the Sun _____

Time needed to rotate on its axis _____

Facts about the planet's surface _____

_____

Facts about the planet's moon(s) _____

_____

Other interesting facts _____

_____

My information came from

_____

_____

A picture of my planet

My planet's symbol

Interesting Moons

Use the code to discover the names of some moons in our solar system.

A	B	C	D	E	F	G	H	I	J	K	L	M
1	2	3	4	5	6	7	8	9	10	11	12	13

N	O	P	Q	R	S	T	U	V	W	X	Y	Z
14	15	16	17	18	19	20	21	22	23	24	25	26

A. Jupiter's moon named ___ ___ has at least eight active volcanoes.
   9   15

B. ___ ___ ___ ___ ___ ___ travels around Mars in $7\frac{1}{2}$ hours. No other
   16 8 15 2 15 19
   moon travels so fast.

C. Jupiter also has the largest moon in the solar system. It is named

   ___ ___ ___ ___ ___ ___ ___ ___.
   7   1   14 25 13 5   4   5

D. ___ ___ ___ ___ ___ is known to have a thick atmosphere. It is one of
   20 9 20 1 14
   Saturn's moons.

E. Neptune's moon ___ ___ ___ ___ ___ ___ orbits the planet backwards.
                    20 18 9 20 15 14

F. ___ ___ ___ ___ ___ ___ is the smallest Martian moon.
   4   5   9   13 15 19

G. ___ ___ ___ ___ ___ ___ is one of Jupiter's 16 moons.
   5   21 18 15 16 1

H. The first footsteps on another surface in space

   were taken on Earth's ___ ___ ___ ___.
                          13 15 15 14

Brainwork! Which moon above would you most like to visit? Write a
paragraph telling which moon you would choose and why.

Beyond Our Solar System

Astronomers know that much lies beyond our solar system. In fact, in the drawing on this page our solar system is just a tiny speck in a larger group of objects in space. This larger group is called the Milky Way galaxy. The Milky Way is made up of all the stars you can see in the night sky and many more beyond those. It also contains large clouds made of gas and dust. But that's not all! Beyond our Milky Way, astronomers have seen millions of other galaxies. Each of these has billions of stars. Astronomers call space and everything in it the universe.

Side View of the Milky Way

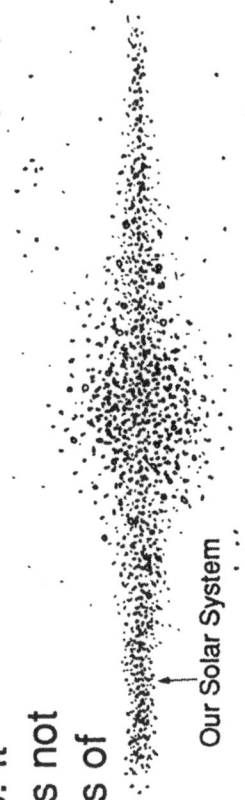

Our Solar System

1. What is the name of our galaxy? _____

2. What have astronomers seen beyond our galaxy? _____

3. What is the universe? _____

4. Which contains the largest group of objects—the solar system, the universe, or the Milky Way? _____

5. What two kinds of objects does the Milky Way contain? _____

Brainwork! Write a mini-book about the universe. Use the words *planet*, *solar system*, and *galaxy*.

19

A Review Riddle

Find a word in the Word Bank that matches each clue below. Write the word on the blanks.

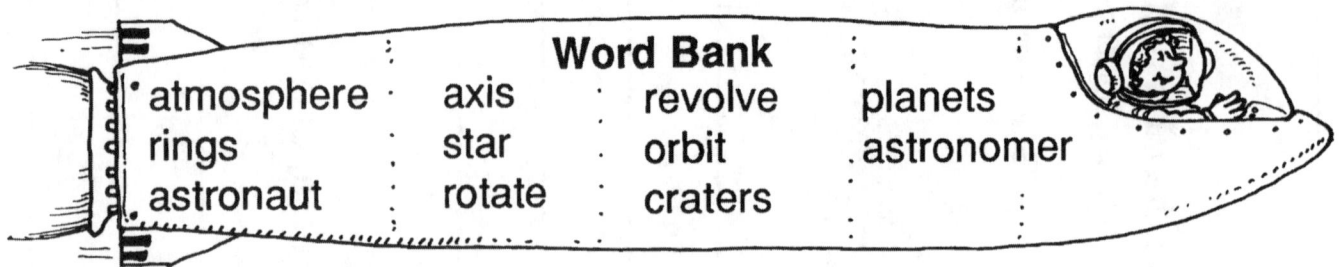

Word Bank

atmosphere   axis   revolve   planets
rings   star   orbit   astronomer
astronaut   rotate   craters

1. person who travels in space ___ ___ ___ ___ ___ ___ (○) ___

2. deep holes (○) ___ ___ ___ ___ ___ ___

3. nine worlds ___ (○) ___ ___ ___ ___ ___

4. to spin (○) ___ ___ ___ ___ ___

5. to travel around ___ ___ ___ ___ ___ (○) ___

6. scientist who studies the objects in space

___ ___ ___ ___ ___ (○) ___ ___ ___ ___

7. imaginary line through the center of a planet ___ ___ (○) ___

8. a planet's path around the sun (○) ___ ___ ___ ___

9. ball of hot glowing gases ___ ___ (○) ___

10. Saturn, Jupiter and Uranus have these ___ ___ ___ (○) ___

11. a blanket of gases ___ ___ ___ ___ ___ (○) ___ ___ ___ ___

Answer this riddle! Write the circled letters on the blanks below.

What is another name for our solar system?

___ ___ ___   ___ ___ ___ ___ ___   ___ ___   ___ ___ ___ ___ ___
8   1   4   11   3   9   2   5   7   6   10   11   9   2   5

Brainwork! Scramble the letters in each planet's name. Have a friend unscramble them.

Answers

Page One
1. a scientist who studies the universe
2. a hot glowing ball of gases
3. nine worlds that travel around the Sun
4. chunks of rock and metal
5. small chunks of rock or ice
6. frozen balls of dirty ice

Page Two
1. A star is a huge glowing ball of gases.
2. The Sun is the only star in our solar system.
3. Even so, the sun is 93 million miles from Earth.
4. The Sun is a yellow star.
5. It looks much larger than the stars we see at night because it is closer to us than the others.

Page Three
1. revolves
2. orbits
3. because the planets are at different distances from the Sun
4. rotates (or spins)
5. axis
6. spins

Page Four
1. Sun
2. Mercury
3. Venus
4. Earth
5. Mars
6. asteroids
7. Jupiter
8. Saturn
9. Uranus
10. Neptune
11. Pluto

Page Six
Suggested answers:
1. Mercury is the planet closest to the Sun.
2. It takes Mercury 88 days to revolve around the Sun.
3. It is hard because of the Sun's great light.
4. Mariner X flew by Mercury three times and sent scientists new information.
5. It is like the moon's surface. It has high cliffs and deep craters.

Page Seven
1. size
2. closest
3. atmosphere
4. light
5. against
6. melt
7. lightning
8. higher

Page Eight
1. Sun
2. planet
3. distance
4. closer
5. liquid
6. reaches
7. soil
8. main
9. Earth

Page Nine
1. craters
2. orbits
3. maria
4. diameter
5. highlands
6. astronaut
Sentences will vary.

Page Ten
1. fourth, two
2. Red Planet
3. 1976, experiments
4. deserts, riverbeds
5. polar caps

Page Eleven
1. false
2. false
3. true
4. true
5. true
6. false
7. true
8. false

Page Twelve
Suggested answers:
1. Saturn is best known for the beautiful rings around its middle.
2. Some scientists believe the rings are made of particles left over from when Saturn first became a planet. OR Others believe the rings are made of pieces of a moon that was torn apart when it came too close to Saturn.
3. Saturn is the second largest planet.
4. Saturn is more than nine times farther than Earth is from the Sun.
5. Saturn has at least 20 moons.

Page Thirteen
1. Uranus
2. both
3. Neptune
4. both
5. both
6. Uranus
7. Uranus
8. Neptune

Page Fourteen
1. Pluto travels farthest from the Sun.
2. Pluto has one moon.
3. Pluto is the smallest planet.
4. Pluto travels more than four billion miles from Earth.
5. Pluto's moon is named Charon.

Page Fifteen
1. 2,788 million miles
2. Mars
3. Mars
4. 31 million miles
5. 49 million miles

Page Sixteen

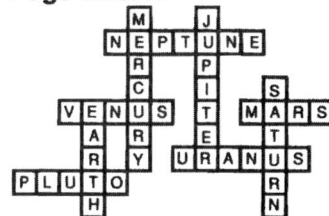

Page Seventeen
Answers will vary.

Page Eighteen
A. IO
B. PHOBOS
C. GANYMEDE
D. TITAN
E. TRITON
F. DEIMOS
G. EUROPA
H. MOON

Page Nineteen
Suggested answers:
1. Our galaxy is the Milky Way.
2. Astronomers have seen millions of other galaxies.
3. The universe is space and everything in it.
4. The universe contains the largest group of objects in space.
5. The Milky Way contains stars and large clouds made of gas and dust.

Page Twenty
1. astronaut
2. craters
3. planets
4. rotate
5. revolve
6. astronomer
7. axis
8. orbit
9. star
10. rings
11. atmosphere
our place in space